SI The International System of Units

Translation approved by the
International Bureau of Weights and Measures
of its publication
'Le Système International d'Unités'

Editors

Chester H. Page
National Bureau of Standards

Paul Vigoureux
National Physical Laboratory

NATIONAL PHYSICAL LABORATORY
Department of Industry

London Her Majesty's Stationery Office

First published 1970
Third edition 1977

Printed in England for
Her Majesty's Stationery Office by
Oyez Press Limited

Dd 496268 K32 6/77

ISBN 0 11 480045 6

Foreword

This document, now published independently by the National Bureau of Standards, USA and Her Majesty's Stationery Office, UK, is a translation of the French 'Le Système International d'Unités' published by the International Bureau of Weights and Measures*. It was prepared jointly by the National Physical Laboratory, UK, and the National Bureau of Standards, USA. The International Bureau of Weights and Measures has compared this translation with the French text and finds that it agrees with the intention and the letter of the original. The International Bureau hopes that wide dissemination of this approved translation will promote knowledge and understanding of the International System of Units, encourage its use in all realms of science, industry and commerce, and secure uniformity of nomenclature throughout the English-speaking world.

J. Terrien DIRECTOR, BIPM

*'Le Système International d'Unités', 1970, OFFILIB, 48 rue Gay-Lussac, F 75005 Paris (Revised edition 1977).

Complete or partial translations of this brochure (or of its earlier editions) have been published in other languages, notably Bulgarian, Czech, German, Japanese, Portuguese, Spanish. Several countries have also published guides for the use of SI units.

Contents

Preface to the 3rd edition

The International Bureau of Weights and Measures (BIPM), in response to frequent requests, publishes this document containing Resolutions and Recommendations of the General Conference of Weights and Measures (CGPM) on the International System of Units. Explanations have been added as well as relevant extracts from the International Standards of the International Organization for Standardization (ISO) for the practical use of the system.

The Consultative Committee for Units (CCU) of the International Committee of Weights and Measures (CIPM) helped to draft the document and has approved the final text.

Appendix I reproduces in chronological order the decisions (Resolutions, Recommendations, Declarations, etc.) promulgated since 1889 by the CGPM and the CIPM on units of measurement and on the International System of Units.

Appendix II outlines the measurements, consistent with the theoretical definitions given here, which metrological laboratories can make to realize the units and to calibrate precision material standards.

This 3rd edition is the second edition brought up to date with the decisions of the 15th CGPM (1975) and the amendments made by the CCU at its 4th and 5th sessions (1974 and 1976), and by the CIPM at its session of September 1976.

January 1977

J. Terrien

DIRECTOR, BIPM

J. de Boer

PRESIDENT, CCU

I Introduction

I.1 Historical note

In 1948 the 9th CGPM*, by its Resolution 6, instructed the CIPM*:

'to study the establishment of a complete set of rules for units of measurement';

'to find out for this purpose, by official enquiry, the opinion prevailing in scientific, technical and educational circles in all countries' and

'to make recommendations on the establishment of a *practical system of units of measurement* suitable for adoption by all signatories to the Metre Convention'.

The same General Conference also laid down, by its Resolution 7, general principles for unit symbols (see II.1.2, page 7) and also gave a list of units with special names.

The 10th CGPM (1954), by its Resolution 6, and the 14th CGPM (1971), by its Resolution 3, adopted as base units of this 'practical system of units', the units of the following seven quantities: length, mass, time, electric current, thermodynamic temperature, amount of substance, and luminous intensity (see II.1, page 3).

The 11th CGPM (1960), by its Resolution 12, adopted the name *International System of Units*, with the international abbreviation SI, for this practical system of units of measurement and laid down rules for the prefixes (see III. 1, page 13), the derived and supplementary units (see II. 2, page 8 and II.3, page 12) and other matters, thus establishing a comprehensive specification for units of measurement.

In the present document the expressions 'SI units', 'SI prefixes', 'supplementary units' are used in accordance with Recommendation 1 (1969) of the CIPM.

* For the meaning of these abbreviations, see the preface.

I.2 The three classes of SI units

SI units are divided into three classes:

base units,

derived units,

supplementary units

From the scientific point of view division of SI units into these three classes is to a certain extent arbitrary, because it is not essential to the physics of the subject.

Nevertheless the General Conference, considering the advantages of a single, practical, world-wide system for international relations, for teaching and for scientific work, decided to base the International System on a choice of seven well-defined units which by convention are regarded as dimensionally independent: the metre, the kilogram, the second, the ampere, the kelvin, the mole, and the candela (see II.1, page 3). These SI units are called *base units*.

The second class of SI units contains *derived units*, i.e. units that can be formed by combining base units according to the algebraic relations linking the corresponding quantities. Several of these algebraic expressions in terms of base units can be replaced by special names and symbols which can themselves be used to form other derived units (see II.2, page 8).

Although it might be thought that SI units can only be base units or derived units, the 11th CGPM (1960) admitted a third class of SI units, called *supplementary units*, for which it declined to state whether they were base units or derived units (see II.3, page 12).

The SI units of these three classes form a coherent set in the sense normally attributed to the expression 'coherent system of units'.

The decimal multiples and sub-multiples of SI units formed by means of SI prefixes must be given their full name *multiples and sub-multiples of SI units* when it is desired to make a distinction between them and the coherent set of SI units.

II SI units

II.1 SI base units

II.1.1 *Definitions*

(a) unit of length (metre)

The 11th CGPM (1960) replaced the definition of the metre based on the international prototype of platinum-iridium, in force since 1889 and amplified in 1927, by the following definition:

The metre is the length equal to 1 650 763.73 *wavelengths in vacuum of the radiation corresponding to the transition between the levels* $2p_{10}$ *and* $5d_5$ *of the krypton-86 atom.*

(11th CGPM (1960), Resolution 6)

The old international prototype of the metre which was legalized by the 1st CGPM in 1889 is still kept at the International Bureau of Weights and Measures under the conditions specified in 1889.

(b) unit of mass (kilogram)

The 1st CGPM (1889) legalized the international prototype of the kilogram and declared: *this prototype shall henceforth be considered to be the unit of mass.*

The 3rd CGPM (1901), in a declaration intended to end the ambiguity which existed as to the meaning of the word 'weight' in popular usage, confirmed that *the kilogram is the unit of mass; it is equal to the mass of the international prototype of the kilogram*, (see the complete declaration, page 22).

This International prototype made of platinum-iridium is kept at the BIPM under conditions specified by the 1st CGPM in 1889.

(c) unit of time (second)

Originally the unit of time, the second, was defined as the fraction 1/86 400 of the mean solar day. The exact definition of 'mean solar day' was left to astronomers, but their measurements have shown that on account of irregularities in the rotation of the Earth the mean solar day does not guarantee the desired accuracy. In order to define the unit of time more precisely the 11th CGPM (1960) adopted a definition given by the International Astronomical Union which was based on the tropical year. Experimental work had however already shown that an atomic standard of time-interval, based on a transition between two energy levels of an atom or a molecule, could be realized and reproduced much more accurately. Considering that a very precise definition of the unit of time of the International System, the second, is indispensable for the needs of advanced metrology, the 13th CGPM (1967) decided to replace the definition of the second by the following:

The second is the duration of 9 192 631 770 *periods of the radiation corresponding to the transition between the two hyperfine levels of the ground state of the caesium -133 atom.*

(13th CGPM (1967), Resolution 1)

(d) unit of electric current (ampere)

Electric units, called 'international', for current and resistance had been introduced by the International Electrical Congress held in Chicago in 1893, and the definitions of the 'international' ampere and the 'international' ohm were confirmed by the International Conference of London in 1908.

Although it was already obvious on the occasion of the 8th CGPM (1933) that there was a unanimous desire to replace those 'international' units by so-called 'absolute' units, the official decision to abolish them was only taken by the 9th CGPM (1948), which adopted for the unit of electric current, the ampere, the following definition:

The ampere is that constant current which, if maintained in two straight parallel conductors of infinite length, of negligible circular cross-section, and placed 1 *metre apart in vacuum, would produce between these conductors a force equal to* 2×10^{-7} *newton per metre of length.*

(CIPM (1946), Resolution 2 approved by the 9th CGPM, 1948)

The expression 'MKS unit of force' which occurs in the original text has been replaced here by 'newton' adopted by the 9th CGPM (1948, Resolution 7).

(e) unit of thermodynamic temperature (kelvin)

The definition of the unit of thermodynamic temperature was given in substance by the 10th CGPM (1954, Resolution 3) which selected the triple point of water as fundamental fixed point and assigned to it the temperature 273.16 K by definition. The 13th CGPM (1967, Resolution 3) adopted the name kelvin (symbol K) instead of 'degree Kelvin' (symbol °K) and in its Resolution 4 defined the unit of thermodynamic temperature as follows:

The kelvin, unit of thermodynamic temperature, is the fraction 1/273.16 *of the thermodynamic temperature of the triple point of water.*

(13th CGPM (1967), Resolution 4)

The 13th CGPM (1967, Resolution 3) also decided that the unit kelvin and its symbol K should be used to express an interval or a difference of temperature.

Note In addition to the thermodynamic temperature (symbol T), expressed in kelvins, use is also made of Celsius temperature (symbol t) defined by the equation
$t = T - T_0$
where $T_0 = 273.15$ K by definition. The unit 'degree Celsius' is equal to the unit 'kelvin', but 'degree Celsius' is a special name in place of 'kelvin' for expressing Celsius temperature. A temperature interval or a Celsius temperature difference can be expressed in degrees Celsius as well as in kelvins.

(f) unit of amount of substance (mole)

Since the discovery of the fundamental laws of chemistry, units of amount of substance called, for instance, 'gram-atom' and 'gram-molecule', have been used to specify amounts of chemical elements or compounds. These units had a direct connection with 'atomic weights' and 'molecular weights', which were in fact relative masses. 'Atomic weights' were originally referred to the atomic weight of oxygen (by general agreement taken as 16). But whereas physicists separated isotopes in the mass spectrograph and attributed the value 16 to one of the isotopes of oxygen, chemists attributed that same value to the (slightly variable) mixture of isotopes 16, 17 and 18 which was for them the naturally occurring element oxygen. Finally an agreement between the International Union of Pure and Applied Physics (IUPAP) and the International Union of Pure and Applied Chemistry (IUPAC) brought this duality to an end in 1959/60. Physicists and chemists have ever since agreed to assign the value 12 to the isotope 12 of carbon. The unified scale thus obtained gives values of 'relative atomic mass'. It remained to define the unit of amount of substance by fixing the corresponding mass of carbon 12; by international agreement this mass has been fixed at 0.012 kg, and the unit of the quantity 'amount of substance'[(1)] has been given the name *mole* (symbol mol).

Following proposals of IUPAP, IUPAC and ISO, the CIPM gave in 1967, and confirmed in 1969, a definition of the mole, eventually adopted by the 14th CGPM (1971, Resolution 3):

1. *The mole is the amount of substance of a system which contains as many elementary entities as there are atoms in* 0.012 *kilogram of carbon* 12.

2. *When the mole is used, the elementary entities must be specified and may be atoms, molecules, ions, electrons, other particles, or specified groups of such particles.*

This definition specifies at the same time the nature of the quantity whose unit is the mole.

(g) unit of luminous intensity (candela)

The units of luminous intensity based on flame or incandescent filament standards in use in various countries were replaced in 1948 by the 'new candle'. This decision had been prepared by the International Commission on Illumination (CIE) and by the CIPM before 1937, and was promulgated by the CIPM at its meeting in 1946 in virtue of powers conferred on it in 1933 by the 8th CGPM. The 9th CGPM (1948) ratified the decision of the CIPM and gave a new international name, candela (symbol cd), to the unit of luminous intensity. The text of the definition of the candela, as amended in 1967, is as follows:

The candela is the luminous intensity, in the perpendicular direction, of a surface of 1/600 000 *square metre of a black body at the temperature of freezing platinum under a pressure of* 101 325 *newtons per square metre.*

(13th CGPM (1967), Resolution 5)

[(1)] The name of this quantity, adopted by IUPAP, IUPAC and ISO, is in French 'quantité de matière', and in English 'amount of substance'; the German and Russian translations are 'Stoffmenge' and 'количество вещества'. The French name recalls 'quantitas materiae' by which in the past the quantity now called mass used to be known; we must forget this old meaning, for mass and amount of substance are entirely different quantities.

II.1.2 *Symbols*

The base units of the International System are collected in Table 1 with their names and their symbols (10th CGPM (1954), Resolution 6; 11th CGPM (1960), Resolution 12; 13th CGPM (1967), Resolution 3; 14th CGPM (1971), Resolution 3).

Table 1 *SI base units*

*Quantity**	*Name*	*Symbol*
length	metre	m
mass	kilogram	kg
time	second	s
electric current	ampere	A
thermodynamic temperature	kelvin	K
amount of substance	mole	mol
luminous intensity	candela	cd

* Translators' note. 'Quantity' is the technical word for measurable attributes of phenomena or matter.

The general principle governing the writing of unit symbols had already been adopted by the 9th CGPM (1948), Resolution 7, according to which:

Roman [upright] *type, in general lower case, is used for symbols of units; if however the symbols are derived from proper names, capital roman type is used* [for the first letter]. *These symbols are not followed by a full stop* [period].

Unit symbols do not change in the plural.

II.2 SI derived units

II.2.1 *Expressions*

Derived units are expressed algebraically in terms of base units by means of the mathematical symbols of multiplication and division. Several derived units have been given special names and symbols which may themselves be used to express other derived units in a simpler way than in terms of the base units.

Derived units may therefore be classified under three headings. Some of them are given in Tables 2, 3 and 4.

Table 2 *Examples of SI derived units expressed in terms of base units*

Quantity	*SI unit*	
	Name	*Symbol*
area	square metre	m^2
volume	cubic metre	m^3
speed, velocity	metre per second	m/s
acceleration	metre per second squared	m/s^2
wave number	1 per metre	m^{-1}
density, mass density	kilogram per cubic metre	kg/m^3
current density	ampere per square metre	A/m^2
magnetic field strength	ampere per metre	A/m
concentration (of amount of substance)	mole per cubic metre	mol/m^3
specific volume	cubic metre per kilogram	m^3/kg
luminance	candela per square metre	cd/m^2

Table 3 *SI derived units with special names*

Quantity	*SI unit*			
	Name	*Symbol*	*Expression in terms of other units*	*Expression in terms of SI base units*
frequency	hertz	Hz		s^{-1}
force	newton	N		$m \cdot kg \cdot s^{-2}$
pressure, stress	pascal	Pa	N/m^2	$m^{-1} \cdot kg \cdot s^{-2}$
energy, work, quantity of heat	joule	J	N·m	$m^2 \cdot kg \cdot s^{-2}$
power, radiant flux	watt	W	J/s	$m^2 \cdot kg \cdot s^{-3}$
quantity of electricity, electric charge	coulomb	C		s·A
electric potential, potential difference, electromotive force	volt	V	W/A	$m^2 \cdot kg \cdot s^{-3} \cdot A^{-1}$
capacitance	farad	F	C/V	$m^{-2} \cdot kg^{-1} \cdot s^4 \cdot A^2$
electric resistance	ohm	Ω	V/A	$m^2 \cdot kg \cdot s^{-3} \cdot A^{-2}$
conductance	siemens	S	A/V	$m^{-2} \cdot kg^{-1} \cdot s^3 \cdot A^2$
magnetic flux	weber	Wb	V·s	$m^2 \cdot kg \cdot s^{-2} \cdot A^{-1}$
magnetic flux density	tesla	T	Wb/m^2	$kg \cdot s^{-2} \cdot A^{-1}$
inductance	henry	H	Wb/A	$m^2 \cdot kg \cdot s^{-2} \cdot A^{-2}$
Celsius temperature[(a)]	degree Celsius	°C		K
luminous flux	lumen	lm		cd·sr[(b)]
illuminance	lux	lx	lm/m^2	$m^{-2} \cdot cd \cdot sr$[(b)]
activity (of a radionuclide)*	becquerel	Bq		s^{-1}
absorbed dose, specific energy imparted, kerma, absorbed dose index	gray	Gy	J/kg	$m^2 \cdot s^{-2}$

(a) See page 5.

(b) In this expression the steradian (sr) is treated as a base unit.

* Translators' note. This term is more appropriate than the direct translation 'ionizing radiations' of the present French text.

Table 4 *Examples of SI derived units expressed by means of special names*

Quantity	*SI unit*		
	Name	*Symbol*	*Expression in terms of SI base units*
dynamic viscosity	pascal second	Pa·s	$m^{-1}·kg·s^{-1}$
moment of force	metre newton	N·m	$m^2·kg·s^{-2}$
surface tension	newton per metre	N/m	$kg·s^{-2}$
power density, heat flux density, irradiance	watt per square metre	W/m^2	$kg·s^{-3}$
heat capacity, entropy	joule per kelvin	J/K	$m^2·kg·s^{-2}·K^{-1}$
specific heat capacity, specific entropy	joule per kilogram kelvin	J/(kg·K)	$m^2·s^{-2}·K^{-1}$
specific energy	joule per kilogram	J/kg	$m^2·s^{-2}$
thermal conductivity	watt per metre kelvin	W/(m·K)	$m·kg·s^{-3}·K^{-1}$
energy density	joule per cubic metre	J/m^3	$m^{-1}·kg·s^{-2}$
electric field strength	volt per metre	V/m	$m·kg·s^{-3}·A^{-1}$
electric charge density	coulomb per cubic metre	C/m^3	$m^{-3}·s·A$
electric flux density	coulomb per square metre	C/m^2	$m^{-2}·s·A$
permittivity	farad per metre	F/m	$m^{-3}·kg^{-1}·s^4·A^2$
permeability	henry per metre	H/m	$m·kg·s^{-2}·A^{-2}$
molar energy	joule per mole	J/mol	$m^2·kg·s^{-2}·mol^{-1}$
molar entropy, molar heat capacity	joule per mole kelvin	J/(mol·K)	$m^2·kg·s^{-2}·K^{-1}·mol^{-1}$
exposure (X and γ rays)	coulomb per kilogram	C/kg	$kg^{-1}·s·A$
absorbed dose rate	gray per second	Gy/s	$m^2·s^{-3}$

Although a derived unit can be expressed in several equivalent ways by using names of base units and special names of derived units, the CIPM sees no objection to the use of certain combinations and certain special names in order to distinguish more easily between quantities of the same dimension. For example the hertz is used, instead of the reciprocal second, for frequency; and the metre-newton, instead of the joule, for the moment of a force. In the field of ionizing radiation, the becquerel is similarly used, instead of the reciprocal second, for activity; and the gray, instead of the joule per kilogram, for specific energy imparted, kerma, absorbed dose, and absorbed dose index.

Note The values of certain so-called dimensionless quantities, as for example refractive index, relative permeability or relative permittivity, are expressed by pure numbers. In this case the corresponding SI unit is the ratio of the same two SI units and may be expressed by the number 1.

II.2.2 *Recommendations*

The International Organization for Standardization (ISO) has issued additional recommendations with the aim of securing uniformity in the use of units, in particular those of the International System (see the series of International Standards ISO 31 and the International Standard ISO 1000 of Technical Committee ISO/TC12 'Quantities, units, symbols, conversion factors and conversion tables').

According to these recommendations:

(*a*) The product of two or more units may be indicated in any of the following ways,

for example: N·m, N.m or N m

(*b*) A solidus (oblique stroke,/), a horizontal line, or negative powers may be used to express a derived unit formed from two others by division

for example: m/s, $\frac{\mathrm{m}}{\mathrm{s}}$ or $\mathrm{m \cdot s^{-1}}$

(*c*) The solidus must not be repeated on the same line unless ambiguity is avoided by parentheses. In complicated cases negative powers or parentheses should be used

for example: $\mathrm{m/s^2}$ or $\mathrm{m \cdot s^{-2}}$ *but not:* m/s/s

$\mathrm{m \cdot kg/(s^3 \cdot A)}$

or $\mathrm{m \cdot kg \cdot s^{-3} \cdot A^{-1}}$ *but not:* $\mathrm{m \cdot kg/s^3/A}$

II.3 SI supplementary units

The General Conference has not yet classified certain units of the International System under either base units or derived units. These SI units are assigned to the third class called 'supplementary units', and may be regarded either as base units or as derived units.

For the time being this class contains only two, purely geometrical, units: the SI unit of plane angle, the *radian*, and the SI unit of solid angle, the *steradian* (11th CGPM (1960), Resolution 12).

Table 5 *SI supplementary units*

Quantity	*SI unit*	
	Name	*Symbol*
plane angle	radian	rad
solid angle	steradian	sr

The radian is the plane angle between two radii of a circle which cut off on the circumference an arc equal in length to the radius.

The steradian is the solid angle which, having its vertex in the centre of a sphere, cuts off an area of the surface of the sphere equal to that of a square with sides of length equal to the radius of the sphere.

(International Standard ISO 31–I)

Supplementary units may be used to form derived units. Examples are given in Table 6.

Table 6 *Examples of SI derived units formed by using supplementary units*

Quantity	*SI unit*	
	Name	*Symbol*
angular velocity	radian per second	rad/s
angular acceleration	radian per second squared	rad/s^2
radiant intensity	watt per steradian	W/sr
radiance	watt per square metre steradian	$W \cdot m^{-2} \cdot sr^{-1}$

III Decimal multiples and sub-multiples of SI units

III.1 SI prefixes

The 11th CGPM (1960, Resolution 12) adopted a first series of names and symbols of prefixes to form decimal multiples and sub-multiples of SI units. Prefixes for 10^{-15} and 10^{-18} were added by the 12th CGPM (1964, Resolution 8), and those for 10^{15} and 10^{18} by the 15th CGPM (1975, Resolution 10).

Table 7 *SI prefixes*

Factor	*Prefix*	*Symbol*	*Factor*	*Prefix*	*Symbol*
10^{18}	exa	E	10^{-1}	deci	d
10^{15}	peta	P	10^{-2}	centi	c
10^{12}	tera	T	10^{-3}	milli	m
10^{9}	giga	G	10^{-6}	micro	μ
10^{6}	mega	M	10^{-9}	nano	n
10^{3}	kilo	k	10^{-12}	pico	p
10^{2}	hecto	h	10^{-15}	femto	f
10^{1}	deca	da	10^{-18}	atto	a

III.2 Recommendations

ISO recommends the following rules for the use of SI prefixes:

(*a*) Prefix symbols are printed in roman (upright) type without spacing between the prefix symbol and the unit symbol.

(*b*) An exponent attached to a symbol containing a prefix indicates that the multiple or sub-multiple of the unit is raised to the power expressed by the exponent,

for example: $1\ \mathrm{cm}^3 = (10^{-2}\ \mathrm{m})^3 = 10^{-6}\ \mathrm{m}^3$
$1\ \mathrm{cm}^{-1} = (10^{-2}\ \mathrm{m})^{-1} = 10^2\ \mathrm{m}^{-1}$
$1\ \mu\mathrm{s}^{-1} = (10^{-6}\ \mathrm{s})^{-1} = 10^6\ \mathrm{s}^{-1}$

(*c*) Compound prefixes, formed by the juxtaposition of two or more SI prefixes, are not to be used,

for example: 1 nm but not 1 mμm

III.3 The kilogram

Among the base units of the International System the unit of mass is the only one whose name, for historical reasons, contains a prefix. Names of decimal multiples and sub-multiples of the unit of mass are formed by attaching prefixes to the word 'gram' (CIPM (1967), Recommendation 2).

IV Units outside the International System

IV.1 Units used with the International System

The CIPM (1969) recognized that users of SI will also wish to employ with it certain units not part of it, but which are important and are widely used. These units are given in Table 8. The combination of units of this table with SI units to form compound units should be restricted to special cases in order not to lose the advantages of the coherence of SI units.

Table 8 *Units in use with the International System*

Name	*Symbol*	*Value in SI unit*
minute	min	1 min = 60 s
hour[(a)]	h	1 hr = 60 min = 3 600 s
day	d	1 d = 24 h = 86 400 s
degree	°	1 ° = $(\pi/180)$ rad
minute	′	1′ = (1/60) ° = $(\pi/10\,800)$ rad
second	″	1″ = (1/60) ′ = $(\pi/648\,000)$ rad
litre[(b)]	l	1 l = 1 dm^3 = 10^{-3} m^3
tonne[(b)]*	t	1 t = 10^3 kg

(a) The symbol of this unit is included in Resolution 7 of the 9th CGPM (1948).

(b) This unit and its symbol were adopted by CIPM in 1879 (Procès-Verbaux CIPM, 1879, p 41). For the symbol for litre, where there is a risk of confusion between the letter l and the number 1, one may use the abbreviation 'ltr' or write 'litre' in full (CIPM, 1976). The present definition of the litre is in Resolution 6 of the 12th CGPM (1964).

* Translator's note. In some English-speaking countries this unit is called 'metric ton'.

It is likewise necessary to recognize, outside the International System, some other units which are useful in specialized fields, because their values expressed in SI units must be obtained by experiment, and are therefore not known exactly (Table 9).

Table 9 *Units, used with the International System, whose values in SI units are obtained experimentally*

Name	*Symbol*	*Definition*
electronvolt	eV	(*a*)
unified atomic mass unit	u	(*b*)
astronomical unit	(*c*)	(*c*)
parsec	pc	(*d*)

(*a*) 1 electronvolt is the kinetic energy acquired by an electron in passing through a potential difference of 1 volt in vacuum; 1 eV $= 1.602\,19 \times 10^{-19}$ J approximately.

(*b*) The unified atomic mass unit is equal to the fraction 1/12 of the mass of an atom of the nuclide ^{12}C; 1 u $= 1.660\,57 \times 10^{-27}$ kg approximately.

(*c*) This unit does not have an international symbol; abbreviations are used, for example, AU in English, UA in French, AE in German, а.е.д. in Russian, etc. The astronomical unit of distance is the length of the radius of the unperturbed circular orbit of a body of negligible mass moving round the Sun with a sidereal angular velocity of 0.017 202 098 950 radian per day of 86 400 ephemeris seconds. In the system of astronomical constants of the International Astronomical Union (1976) the value adopted is:
1 AU $= 149\,597.870 \times 10^6$ m.

(*d*) 1 parsec is the distance at which 1 astronomical unit subtends an angle of 1 second of arc; we thus have approximately, 1 pc = 206 265 AU = $30\,857 \times 10^{12}$ m.

IV.2 Units accepted temporarily

In view of existing practice the CIPM (1969) considered it was preferable to keep temporarily, for use with those of the International System, the units listed in Table 10.

Table 10 *Units to be used temporarily with the International System*

Name	*Symbol*	*Value in SI units*
nautical mile[a]		1 nautical mile = 1852 m
knot		1 nautical mile per hour = (1852/3600) m/s
ångström	Å	1 Å = 0.1 nm = 10^{-10} m
are [b]	a	1 a = 1 dam^2 = 10^2 m^2
hectare[b]	ha	1 ha = 1 hm^2 = 10^4 m^2
barn[c]	b	1 b = 100 fm^2 = 10^{-28} m^2
bar[d]	bar	1 bar = 0.1 MPa = 10^5 Pa
standard atmosphere[e]	atm	1 atm = 101 325 Pa
gal[f]	Gal	1 Gal = 1 cm/s^2 = 10^{-2} m/s^2
curie[g]	Ci	1 Ci = 3.7×10^{10} Bq
röntgen[h]	R	1 R = 2.58×10^{-4} C/kg
rad[i]	rad	1 rad = 1 cGy = 10^{-2} Gy

[a] The nautical mile is a special unit employed for marine and aerial navigation to express distances. The conventional value given above was adopted by the First International Extraordinary Hydrographic Conference, Monaco, 1929, under the name 'International nautical mile'.

[b] This unit and its symbol were adopted by the CIPM in 1879 (*Procès-Verbaux CIPM*, 1879, p. 41).

[c] The barn is a special unit employed in nuclear physics to express effective cross-sections.

[d] This unit and its symbol are included in Resolution 7 of the 9th CGPM (1948).

[e] Resolution 4 of 10th CGPM (1954).

[f] The gal is a special unit employed in geodesy and geophysics to express the acceleration due to gravity.

[g] The curie is a special unit employed in nuclear physics to express activity of radionuclides (12th CGPM (1964), Resolution 7).

[h] The röntgen is a special unit employed to express exposure of X or γ radiations.

[i] The rad is a special unit employed to express absorbed dose of ionizing radiations. When there is risk of confusion with the symbol for radian, rd may be used as symbol for rad.

IV.3 CGS units

The CIPM considers that it is in general preferable not to use, with the units of the International System, CGS units which have special names[(2)]. Such units are listed in Table 11.

Table 11 *CGS units with special names*

Name	*Symbol*	*Value in SI units*
erg[(a)]	erg	1 erg $= 10^{-7}$ J
dyne[(a)]	dyn	1 dyn $= 10^{-5}$ N
poise[(a)]	P	1 P = 1 dyn·s/cm² = 0.1 Pa·s
stokes	St	1 St = 1 cm²/s $= 10^{-4}$ m²/s
gauss[(b)]	Gs,G	1 Gs corresponds to 10^{-4} T
oersted[(b)]	Oe	1 Oe corresponds to $(1000/4\pi)$ A/m
maxwell[(b)]	Mx	1 Mx corresponds to 10^{-8} Wb
stilb[(a)]	sb	1 sb = 1 cd/cm² = 10^4 cd/m²
phot	ph	1 ph = 10^4 lx

[(a)] This unit and its symbol were included in Resolution 7 of the 9th CGPM (1948).

[(b)] This unit is part of the so-called 'electromagnetic' 3-dimensional CGS system and cannot strictly speaking be compared to the corresponding unit of the International System, which has four dimensions when only electric quantities are considered.

[(2)] The aim of the International System of Units and of the recommendations contained in this document is to secure a greater degree of uniformity, hence a better mutual understanding of the general use of units. Nevertheless in certain specialized fields of scientific research, in particular in theoretical physics, there may sometimes be very good reasons for using other systems or other units.

Whichever units are used, it is important that the *symbols* employed for them follow current international recommendations.

IV.4 Other units

As regards units outside the International System which do not come under Sections IV.1, 2 and 3, the CIPM considers that it is in general preferable to avoid them, and to use instead units of the International System. Some of those units are listed in Table 12.

Table 12 *Other units generally deprecated*

Name	*Value in SI units*
fermi	1 fermi = 1 fm = 10^{-15} m
metric carat[a]	1 metric carat = 200 mg = 2×10^{-4} kg
torr	1 torr = (101 325/760) Pa
kilogram-force (kgf)	1 kgf = 9.806 65 N
calorie (cal)	1 cal = 4.186 8 J[b]
micron (μ)[c]	1μ = 1μm = 10^{-6} m
X unit[d]	
stere (st)[e]	1 st = 1 m^3
gamma (γ)	1 γ = 1 nT = 10^{-9} T
γ[f]	1 γ = 1 μg = 10^{-9} kg
λ[g]	1 λ = 1 μl = 10^{-6} l = 10^{-9} m^3

(a) This name was adopted by the 4th CGPM (1907, pp. 89–91) for commercial dealings in diamonds, pearls and precious stones.

(b) This value is that of the so-called 'IT' calorie (5th International Conference on Properties of Steam, London, 1956).

(c) The name of this unit and its symbol, adopted by the CIPM in 1879 (*Procès-Verbaux CIPM*, 1879, p. 41) and repeated in Resolution 7 of the 9th CGPM (1948) were abolished by the 13th CGPM (1967, Resolution 7).

(d) This special unit was employed to express wavelengths of X rays; 1 X unit = 1.002×10^{-4} nm approximately.

(e) This special unit employed to measure firewood was adopted by the CIPM in 1879 with the symbol 's' (*Procès-Verbaux CIPM*, 1879, p. 41). The 9th CGPM (1948, Resolution 7) changed the symbol to 'st'.

(f) This symbol is mentioned in *Procès-Verbaux CIPM*, 1880, p. 56.

(g) This symbol is mentioned in *Procès-Verbaux CIPM*, 1880, p. 30.

Appendix I Decisions of the CGPM and CIPM

(The more important decisions abrogated, modified, or added to, are indicated by an asterisk*)

CR: *Comptes rendus des séances de la Conférence Générale des Poids et Mesures (CGPM)*

PV: *Procès-Verbaux des séances du Comité International des Poids et Mesures (CIPM)*

1st CGPM, 1889

metre
kilogram

Sanction of the international prototypes of the metre and the kilogram (CR, 34–38)

The General Conference

considering

the 'Compte rendu of the President of the CIPM' and the 'Report of the CIPM', which show that, by the collaboration of the French section of the international Metre Commission and of the CIPM, the fundamental measurements of the international and national prototypes of the metre and of the kilogram have been made with all the accuracy and reliability which the present state of science permits;

that the international and national prototypes of the metre and the kilogram are made of an alloy of platinum with 10 per cent iridium, to within 0.0001;

the equality in length of the international Metre and the equality in mass of the international Kilogram with the length of the Metre and the mass of the Kilogram kept in the Archives of France;

that the differences between the national Metres and the international Metre lie within 0.01 millimetre and that these differences are based on a hydrogen thermometer scale which can always be reproduced thanks to the stability of hydrogen, provided identical conditions are secured;

that the differences between the national Kilograms and the international Kilogram lie within 1 milligram;

that the international Metre and Kilogram and the national Metres and Kilograms fulfil the requirements of the Metre Convention, *sanctions*

A. As regards international prototypes:

1. The Prototype of the metre chosen by the CIPM.

This prototype, at the temperature of melting ice, shall henceforth represent the metric unit of length.*

2. The Prototype of the Kilogram adopted by the CIPM.
This prototype shall henceforth be considered as the unit of mass.

3. The hydrogen thermometer centigrade scale in terms of which the equations of the prototype Metres have been established.

B. As regards national prototypes:

. . . .

* Definition abrogated in 1960 (see p. 29: 11th CGPM, Resolution 6).

3rd CGPM, 1901

litre

Declaration concerning the definition of the litre (CR, 38)

. . . .

The Conference declares:

1. The unit of volume, for high accuracy determinations, is the volume occupied by a mass of 1 kilogram of pure water, at its maximum density and at standard atmospheric pressure; this volume is called 'litre'.*

2. . . .

* Definition abrogated in 1964 (see p. 34: 12th CGPM. Resolution 6).

mass and weight
g_n

Declaration on the unit of mass and on the definition of weight; conventional value of g_n (CR, 70)

Taking into account the decision of the CIPM of the 15 October 1887, according to which the kilogram has been defined as unit of mass(1);

Taking into account the decision contained in the sanction of the prototypes of the Metric System, unanimously accepted by the CGPM on the 26 September 1889;

Considering the necessity to put an end to the ambiguity which in current practice still subsists on the meaning of the word *weight*, used sometimes for *mass*, sometimes for *mechanical force*;

The Conference declares:

'1. The kilogram is the unit of mass; it is equal to the mass of the international prototype of the kilogram;

(1) 'The mass of the international Kilogram is taken as unit for the International Service of Weights and Measures' (PV, 1887, 88).

'2. The word *weight* denotes a quantity of the same nature as a *force*; the weight of a body is the product of its mass and the acceleration due to gravity; in particular, the standard weight of a body is the product of its mass and the standard acceleration due to gravity;

'3. The value adopted in the International Service of Weights and Measures for the standard acceleration due to gravity is 980.665 cm/s^2, value already stated in the laws of some countries.'(2)

7th CGPM, 1927

metre *Definition of the metre by the international Prototype* (CR, 49)

The unit of length is the metre, defined by the distance, at 0°, between the axes of the two central lines marked on the bar of platinum-iridium kept at the BIPM and declared Prototype of the metre by the 1st CGPM, this bar being subject to standard atmospheric pressure and supported on two cylinders of at least one centimetre diameter, symmetrically placed in the same horizontal plane at a distance of 571 mm from each other.*

* Definition abrogated in 1960 (see p. 29: 11th CGPM, Resolution 6).

CIPM, 1946

photometric units *Definitions of photometric units* (PV, **20**, 119)

Resolution(3)

. . . .

4. The photometric units may be defined as follows:

New candle (unit of luminous intensity) The value of the new candle is such that the brightness of the full radiator at the temperature of solidification of platinum is 60 new candles per square centimetre.*

New lumen (unit of luminous flux) The new lumen is the luminous flux emitted in unit solid angle (steradian) by a uniform point source having a luminous intensity of 1 new candle.

5. . . .

* Definition modified in 1967 (see p. 36: 13th CGPM, Resolution 5).

(2) *Note of BIPM.* This conventional reference 'standard value' (g_n = 9.80665 m/s^2) to be used in the reduction to standard gravity of measurements made in some place on the Earth has been reconfirmed in 1913 by the 5th CGPM (CR, 44).

(3) The two definitions contained in this Resolution were ratified by the 9th CGPM (1948), which also approved the name *candela* given to the 'new candle' (CR, 54). For the lumen the qualifier 'new' was later abandoned.

mechanical and electric units

Definitions of electric units (PV, **20**, 131)

Resolution 2[(4)]

. . . .

4. (A) Definitions of the mechanical units which enter the definitions of electric units:

Unit of force The unit of force [in the MKS (Metre, Kilogram, Second) system] is the force which gives to a mass of 1 kilogram an acceleration of 1 metre per second, per second*

* The name 'newton' was adopted in 1948 for the MKS unit of force (see Note 4).

Joule (unit of energy or work) The joule is the work done when the point of application of 1 MKS unit of force [newton] moves a distance of 1 metre in the direction of the force.

Watt (unit of power) The watt is the power which in one second gives rise to energy of 1 joule.

(B) Definitions of electric units. The CIPM accepts the following propositions which define the theoretical value of the electric units:

Ampere (unit of electric current) The ampere is that constant current which, if maintained in two straight parallel conductors of infinite length, of negligible circular cross-section, and placed 1 metre apart in vacuum, would produce between these conductors a force equal to 2×10^{-7} MKS unit of force [newton] per metre of length.

Volt (unit of potential difference and of electromotive force) The volt is the potential difference between two points of a conducting wire carrying a constant current of 1 ampere, when the power dissipated between these points is equal to 1 watt.

Ohm (unit of electric resistance) The ohm is the electric resistance between two points of a conductor when a constant potential difference of 1 volt, applied to these points, produces in the conductor a current of 1 ampere, the conductor not being the seat of any electromotive force.

Coulomb (unit of quantity of electricity) The coulomb is the quantity of electricity carried in 1 second by a current of 1 ampere.

Farad (unit of capacitance) The farad is the capacitance of a capacitor between the plates of which there appears a potential difference of 1 volt when it is charged by a quantity of electricity of 1 coulomb.

Henry (unit of electric inductance) The henry is the inductance of a closed circuit in which an electromotive force of 1 volt is produced when the electric current in the circuit varies uniformly at the rate of 1 ampere per second.

[(4)] The definitions contained in this Resolution 2 were approved by the 9th CGPM (1948), (CR, 49), which moreover adopted the name *newton* (Resolution 7).

Weber (unit of magnetic flux) The weber is the magnetic flux which, linking a circuit of one turn, would produce in it an electromotive force of 1 volt if it were reduced to zero at a uniform rate in 1 second.

9th CGPM, 1948

thermodynamic scale, unit of quantity of heat

Triple point of water; thermodynamic scale with a single fixed point; unit of quantity of heat (joule) (CR, 55 and 63)

Resolution 3[5]

1. With present-day technique, the triple point of water is capable of providing a thermometric reference point with an accuracy higher than can be obtained from the melting point of ice.

In consequence the Consultative Committee [for Thermometry and Calorimetry] considers that the zero of the centesimal thermodynamic scale must be defined as the temperature 0.0100 degree below that of the triple point of water.

2. The CCTC accepts the principle of an absolute thermodynamic scale with a single fundamental fixed point, at present provided by the triple point of pure water, the absolute temperature of which will be fixed at a later date.

The introduction of this new scale does not affect in any way the use of the International Scale, which remains the recommended practical scale.

3. The unit of quantity of heat is the joule.

Note It is requested that the results of calorimetric experiments be as far as possible expressed in joules.

If the experiments are made by comparison with the rise of temperature of water (and that, for some reason, it is not possible to avoid using the calorie), the information necessary for conversion to joules must be provided.

The CIPM, advised by the CCTC, should prepare a table giving, in joules per degree, the most accurate values that can be obtained from experiments on the specific heat of water.

degree Celsius

Adoption of 'degree Celsius'

From three names ('degree centigrade', 'centesimal degree', 'degree Celsius') proposed to denote the degree of temperature, the CIPM has chosen 'degree Celsius' (PV, **21**, 1948, 88).

This name is also adopted by the General Conference (CR, 64).

[5] The three propositions contained in this Resolution 3 have been adopted by the General Conference.

practical system of units of measurement

Proposal for establishing a practical system of units of measurement (CR, 64)

Resolution 6
The General Conference,

considering

that the CIPM has been requested by the International Union of Physics to adopt for international use a practical international system of units; that the International Union of Physics recommends the MKS system and one electric unit of the absolute practical system, but does not recommend that the CGS system be abandoned by physicists;

that the CGPM has itself received from the French Government a similar request, accompanied by a draft to be used as basis of discussion for the establishment of a complete specification of units of measurement;

instructs the CIPM:

to seek by an energetic, active, official enquiry the opinion of scientific, technical and educational circles of all countries (offering them in effect the French document as basis);

to gather and study the answers;

to make recommendations for a single practical system of units of measurement, suitable for adoption by all countries adhering to the Metre Convention.

symbols and numbers

Writing and printing of unit symbols and of numbers (CR, 70)

Resolution 7

Principles

Roman (upright) type, in general lower case, is used for symbols of units; if however the symbols are derived from proper names, capital roman type is used. These symbols are not followed by a full stop.

In numbers, the comma (French practice) or the dot (British practice) is used only to separate the integral part of numbers from the decimal part. Numbers may be divided in groups of three in order to facilitate reading; neither dots nor commas are ever inserted in the spaces between groups.

Unit	*Symbol*	*Unit*	*Symbol*
·metre	m	ampere	A
·square metre	m^2	volt	V
·cubic metre	m^3	watt	W
·micron*	μ	ohm	Ω
·litre	l	coulomb	C
·gram	g	farad	F
·tonne	t	henry	H
second	s	hertz	Hz
erg	erg	poise	P
dyne	dyn	newton	N
degree Celsius	°C	·candela ('new candle')	cd
·degree absolute**	°K	lux	lx
calorie	cal	lumen	lm
bar	bar	stilb	sb
hour	h		

* Unit and symbol abrogated in 1967 (see p. 37: 13th CGPM, Resolution 7).

** Name and symbol changed in 1967 (see p. 35: 13th CGPM, Resolution 3).

Notes

1. The symbols whose unit names are preceded by dots are those which had already been adopted by a decision of the CIPM.

2. The symbol for the stere, the unit of volume for firewood, shall be 'st' and not 's', which had been previously assigned to it by the CIPM.

3. To indicate a temperature interval or difference, rather than a temperature, the word 'degree' in full, or the abbreviation 'deg', must be used*.

* See p. 35, Resolution 3 of the 13th CGPM, 1967.

10th CGPM, 1954

thermodynamic scale

Definition of the thermodynamic temperature scale (CR, 79)

Resolution 3

The 10th CGPM decided to define the thermodynamic temperature scale by choosing the triple point of water as fundamental fixed point, and assigning to it the temperature 273.16 degrees Kelvin, exactly.*

* See p. 36, Resolution 4 of the 13th CGPM, 1967.

standard atmosphere

Definition of standard atmosphere (CR, 79)

Resolution 4

The 10th CGPM, having noted that the definition of the standard atmosphere given by the 9th CGPM when defining the International

Temperature Scale, led some physicists to believe that this definition of the standard atmosphere was valid only for accurate work in thermometry,

declares that it adopts, for general use, the definition:

1 standard atmosphere = 1 013 250 dynes per square centimetre,

i.e. 101 325 newtons per square metre.

practical system of units

Practical system of units (CR, 80)

Resolution 6

In accordance with the wish expressed by the 9th CGPM in its Resolution 6 concerning the establishment of a practical system of units of measurement for international use, the 10th CGPM

decides to adopt as base units of the system, the following units:

length	metre
mass	kilogram
time	second
electric current	ampere
thermodynamic temperature	degree Kelvin*
luminous intensity	candela

* Name changed to 'kelvin' in 1967 (see p. 35: 13th CGPM, Resolution 3).

CIPM, 1956

second

Definition of the unit of time (PV, **25,** 77)

Resolution 1

In virtue of the powers invested in it by Resolution 5 of the 10th CGPM,

considering

1. that the 9th General Assembly of the International Astronomical Union (Dublin, 1955) declared itself in favour of linking the second to the tropical year;

2. that, according to the decisions of the 8th General Assembly of the International Astronomical Union (Rome, 1952) the second of ephemeris time (ET) is the fraction $\frac{12\,960\,276\,813}{408\,986\,496} \times 10^{-9}$ of the tropical year for 1900 January 0 at 12 h ET,

decides

'The second is the fraction 1/31 556 925.974 7 of the tropical year for 1900 January 0 at 12 hours ephemeris time'.*

* Definition abrogated in 1967 (see p. 35: 13th CGPM, Resolution 1).

SI *International System of Units* (PV, **25**, 83)

Resolution 3
The CIPM

considering

the task entrusted to it by Resolution 6 of the 9th CGPM concerning the establishment of a practical system of units of measurement suitable for adoption by all countries adhering to the Metre Convention,

the documents received from twenty-one countries in reply to the enquiry requested by the 9th CGPM,

Resolution 6 of the 10th CGPM, fixing the base units of the system to be established

recommends

1. that the name 'International System of Units' be given to the system founded on the base units adopted by the 10th CGPM, viz:
(here follows the list of the six base units with their symbols, reproduced in Resolution 12 of the 11th CGPM (1960)).

2. that the units listed in the table below be used, without excluding others which might be added later: (here follows the table of units reproduced in paragraph 4 of Resolution 12 of the 11th CGPM (1960)).

11th CGPM, 1960

metre *Definition of the metre* (CR, 85)

Resolution 6
The 11th CGPM

considering

that the international Prototype does not define the metre with an accuracy adequate for the present needs of metrology,

that it is moreover desirable to adopt a natural and indestructible standard,

decides,

1. The metre is the length equal to 1 650 763.73 wavelengths in vacuum of the radiation corresponding to the transition between the levels $2p_{10}$ and $5d_5$ of the krypton-86 atom.

2. The definition of the metre in force since 1889, based on the international Prototype of platinum-iridium, is abrogated.

3. The international Prototype of the metre sanctioned by the 1st CGPM in 1889 shall be kept at the BIPM under the conditions specified in 1889.

Resolution 7
The 11th CGPM

requests the CIPM

1. to prepare specifications for the realization of the new definition of the metre[(6)];

2. to select secondary wavelength standards for measurement of length by interferometry, and to prepare specifications for their use;

3. to continue the work in progress on improvement of length standards.

second ***Definition of the unit of time*** (CR, 86)

Resolution 9
The 11th CGPM

considering

the powers given to the CIPM by the 10th CGPM to define the fundamental unit of time,

the decision taken by the CIPM in 1956,

ratifies the following definition:

'The second is the fraction 1/31 556 925.974 7 of the tropical year for 1900 January 0 at 12 hours ephemeris time'.*

* Definition abrogated in 1967 (see p. 35: 13th CGPM, Resolution 1).

SI ***International System of Units*** (CR, 87)

Resolution 12
The 11th CGPM

considering

Resolution 6 of the 10th CGPM, by which it adopted six base units on which to establish a practical system of measurement for international use:

length	metre	m
mass	kilogram	kg
time	second	s
electric current	ampere	A
thermodynamic temperature	degree kelvin	°K*
luminous intensity	candela	cd

* Name and symbol of unit modified in 1967 (see p. 35: 13th CGPM, Resolution 3).

(6) See Appendix 2, page 43, for the relevant Recommendation adopted by the CIPM.

Resolution 3 adopted by the CIPM in 1956,

the recommendations adopted by the CIPM in 1958 concerning an abbreviation for the name of the system, and prefixes to form multiples and sub-multiples of the units,

decides

1. the system founded on the six base units above is called 'International System of Units';*

2. the international abbreviation of the name of the system is: SI;

3. names of multiples and sub-multiples of the units are formed by means of the following prefixes:**

Multiplying factor	*Prefix*	*Symbol*
1 000 000 000 000 = 10^{12}	tera	T
1 000 000 000 = 10^{9}	giga	G
1 000 000 = 10^{6}	mega	M
1 000 = 10^{3}	kilo	k
100 = 10^{2}	hecto	h
10 = 10^{1}	deca	da
0.1 = 10^{-1}	deci	d
0.01 = 10^{-2}	centi	c
0.001 = 10^{-3}	milli	m
0.000 001 = 10^{-6}	micro	μ
0.000 000 001 = 10^{-9}	nano	n
0.000 000 000 001 = 10^{-12}	pico	p

* A seventh base unit, the mole, was adopted in 1971 by the 14th CGPM (Resolution 3, see p. 40).

** See pages 34 and 41 for the four new prefixes adopted by the 12th CGPM (1964), Resolution 8, and the 15th CGPM (1975), Resolution 10.

4. the units listed below are used in the system, without excluding others which might be added later

Supplementary units

plane angle	radian	rad	
solid angle	steradian	sr	

*Derived units**

area	square metre	m^2	
volume	cubic metre	m^3	
frequency	hertz	Hz	1/s
mass density (density)	kilogram per cubic metre	kg/m^3	
speed, velocity	metre per second	m/s	
angular velocity	radian per second	rad/sec	
acceleration	metre per second squared	m/s^2	
angular acceleration	radian per second squared	rad/s^2	
force	newton	N	$kg \cdot m/s^2$
pressure (mechanical stress)	newton per square metre	N/m^2	
kinematic viscosity	square metre per second	m^2/s	
dynamic viscosity	newton-second per square metre	$N \cdot s/m^2$	
work, energy, quantity of heat	joule	J	N·m
power	watt	W	J/s
quantity of electricity	coulomb	C	A·s
potential difference electromotive force	volt	V	W/A
electric field strength	volt per metre	V/m	
electric resistance	ohm	Ω	V/A
capacitance	farad	F	A·s/V
magnetic flux	weber	Wb	V·s
inductance	henry	H	V·s/A
magnetic flux density	tesla	T	Wb/m^2
magnetic field strength	ampere per metre	A/m	
magnetomotive force	ampere	A	
luminous flux	lumen	lm	cd·sr
luminance	candela per square metre	cd/m^2	
illuminance	lux	lx	lm/m^2

* see p. 37 for the other units added by the 13th CGPM (1967), Resolution 6.

cubic decimetre and litre

Cubic decimetre and litre (CR, 88)

Resolution 13
The 11th CGPM,

considering

that the cubic decimetre and the litre are unequal and differ by about 28 parts in 10^6, that determinations of physical quantities which involve measurements of volume are being made more and more accurately, thus increasing the risk of confusion between the cubic decimetre and the litre,

requests the CIPM to study the problem and submit its conclusions to the 12th CGPM.

CIPM, 1961

Cubic decimetre and litre (PV, **29,** 34)

Recommendation

the CIPM recommends that the results of accurate measurements of volume be expressed in units of the International System and not in litres.

12th CGPM, 1964

frequency standard

Atomic standard of frequency (CR, 93)

Resolution 5
The 12th CGPM,

considering

that the 11th CGPM noted in its Resolution 10 the urgency, in the interests of accurate metrology, of adopting an atomic or molecular standard of time interval,

that, in spite of the results already obtained with caesium atomic frequency standards, the time has not yet come for the CGPM to adopt a new definition of the second, base unit of the International System of Units, because of the new and considerable improvements likely to be obtained from work now in progress,

considering also that it is not desirable to wait any longer before time measurements in physics are based on atomic or molecular frequency standards,

empowers the CIPM to name the atomic or molecular frequency standards to be employed for the time being,

requests the Organizations and Laboratories knowledgeable in this field to pursue work connected with a new definition of the second.

Declaration of the CIPM (1964) (PV, **32**, 26 and CR, 93)

The CIPM,

empowered by Resolution 5 of the 12th CGPM to name atomic or molecular frequency standards for temporary use for time measurements in physics, *declares* that the standard to be employed is the transition between the hyperfine levels F = 4, M = 0 and F = 3, M = 0 of the ground state $^2S_{1/2}$ of the caesium-133 atom, unperturbed by external fields, and that the frequency of this transition is assigned the value 9 192 631 770 hertz.

litre *Litre* (CR, 93)

Resolution 6
The 12th CGPM,

considering Resolution 13 adopted by the 11th CGPM in 1960 and the Recommendation adopted by the CIPM in 1961,

1. *abrogates* the definition of the litre given in 1901 by the 3rd CGPM,

2. *declares* that the word 'litre' may be employed as a special name for the cubic decimetre,

3. *recommends* that the name litre should not be employed to give the results of high accuracy volume measurements.

curie *Curie* (CR, 94)

Resolution 7
The 12th CGPM,

considering that the curie has been used for a long time in many countries as unit of activity for radionuclides,

recognizing that in the International System of Units (SI), the unit of this activity is the second to the power of minus one (s^{-1}),*

accepts that the curie be still retained, outside SI, as unit of activity, with the value 3.7×10^{10} s^{-1}. The symbol for this unit is Ci.

* In 1975 the name 'becquerel' (Bq) was adopted for the SI unit of activity (see p. 41: 15th CGPM, Resolution 8); 1 Ci = 3.7×10^{10} Bq.

femto and atto ***SI prefixes femto and atto*** (CR, 94)

Resolution 8
The 12th CGPM,

decides to add to the list of prefixes for the formation of names of multiples and sub-multiples of units, adopted by the 11th CGPM, Resolution 12, paragraph 3, the following two new prefixes:

Multiplying factor	*Prefix*	*Symbol*
10^{-15}	femto	f
10^{-18}	atto	a

second *SI unit of time (second)* (CR, 103)

Resolution 1
The 13th CGPM,

considering

that the definition of the second adopted by the CIPM in 1956 (Resolution 1) and ratified by Resolution 9 of the 11th CGPM (1960), later upheld by Resolution 5 of the 12th CGPM (1964), is inadequate for the present needs of metrology,

that at its meeting of 1964 the CIPM, empowered by Resolution 5 of the 12th CGPM (1964) recommended, in order to fulfil these requirements, a caesium atomic frequency standard for temporary use,

that this frequency standard has now been sufficiently tested and found sufficiently accurate to provide a definition of the second fulfilling present requirements,

that the time has now come to replace the definition now in force of the unit of time of the International System of Units by an atomic definition based on that standard,

decides

1. The unit of time of the International System of Units is the second defined as follows:
'The second is the duration of 9 192 631 770 periods of the radiation corresponding to the transition between the two hyperfine levels of the ground state of the caesium-133 atom'.

2. Resolution 1 adopted by the CIPM at its meeting of 1956 and Resolution 9 of the 11th CGPM are now abrogated.

kelvin (degree Celsius) *SI unit of thermodynamic temperature (kelvin)* (CR, 104)

Resolution 3
The 13th CGPM,

considering

the names 'degree Kelvin' and 'degree', the symbols '°K' and 'deg' and the rules for their use given in Resolution 7 of the 9th CGPM (1948), in Resolution 12 of the 11th CGPM (1960), and the decision taken by the CIPM in 1962 (PV, **30**, 27) (7),

(7) '1. The unit degree Kelvin (symbol °K) may be employed for a difference of two thermodynamic temperatures as well as for thermodynamic temperature itself.

2. If it is found necessary to omit the name Kelvin, the international symbol 'deg' is recommended for the unit of difference of temperature. (The symbol 'deg' is read, for example: 'degré' in French, 'degree' in English, 'gradous' (градус) in Russian, 'Grad' in German, 'graad' in Dutch)'.

that the unit of thermodynamic temperature and the unit of temperature interval are one and the same unit, which ought to be denoted by a single name and a single symbol,

decides

1. the unit of thermodynamic temperature is denoted by the name 'kelvin' and its symbol is 'K';

2. the same name and the same symbol are used to express a temperature interval;

3. a temperature interval may also be expressed in degrees Celsius;

4. the decisions mentioned in the opening paragraph concerning the name of the unit of thermodynamic temperature, its symbol and the designation of the unit to express an interval or a difference of temperature are abrogated, but the usages which derive from these decisions remain permissible for the time being.

kelvin Resolution 4
The 13th CGPM,

considering that it is useful to formulate more explicitly the definition of the unit of thermodynamic temperature contained in Resolution 3 of the 10th CGPM (1954),

decides to express this definition as follows:

'The kelvin, unit of thermodynamic temperature, is the fraction 1/273.16 of the thermodynamic temperature of the triple point of water'.

candela ***SI unit of luminous intensity (candela)***(CR, 104)

Resolution 5
The 13th CGPM,

considering

the definition of the unit of luminous intensity ratified by the 9th CGPM (1948) and contained in the 'Resolution concerning the change of photometric units' adopted by the CIPM in 1946 (PV, **20,** 119) in virtue of the powers conferred by the 8th CGPM (1933),

that this definition fixes satisfactorily the unit of luminous intensity, but that its wording may be open to criticism,

decides to express the definition of the candela as follows:

'The candela is the luminous intensity, in the perpendicular direction, of a surface of 1/600 000 square metre of a black body at the temperature of freezing platinum under a pressure of 101 325 newtons per square metre'.

SI derived units *SI derived units* (CR, 105)

Resolution 6
The 13th CGPM

considering that it is useful to add some derived units to the list of paragraph 4 of Resolution 12 of the 11th CGPM (1960),

decides to add:

wave number	1 per metre	m^{-1}
entropy	joule per kelvin	J/K
specific heat capacity	joule per kilogram kelvin	J/(kg·K)
thermal conductivity	watt per metre kelvin	W/(m·K)
radiant intensity	watt per steradian	W/sr
activity (of a radioactive source)	1 per second	s^{-1}*

*Name and symbol of the unit changed in 1975 (see p. 41: 15th CGPM, Resolution 8).

micron (μ) new candle *Abrogation of earlier decisions (micron, new candle)* (CR, 105)

Resolution 7
The 13th CGPM,

considering that subsequent decisions of the General Conference concerning the International System of Units are incompatible with parts of Resolution 7 of the 9th CGPM (1948),

decides accordingly to remove from Resolution 7 of the 9th Conference:

1. the unit name 'micron', and the symbol 'μ' which had been given to that unit, but which has now become a prefix;

2. the unit name 'new candle'.

CIPM, 1967

multiples of kilogram *Decimal multiples and sub-multiples of the unit of mass* (PV, **35,** 29)

Recommendation 2
The CIPM,

considering that the rule for forming names of decimal multiples and sub-multiples of the units of paragraph 3 of Resolution 12 of the 11th CGPM (1960) might be interpreted in different ways when applied to the unit of mass,

declares that the rules of Resolution 12 of the 11th CGPM apply to the kilogram in the following manner: the names of decimal multiples and sub-multiples of the unit of mass are formed by attaching prefixes to the word 'gram'.

CIPM, 1969

SI ***International System of Units: Rules for application of Resolution 12 of the 11th CGPM (1960)*** (PV, 37, 30)

Recommendation 1 (1969)
The CIPM,

considering that Resolution 12 of the 11th CGPM (1960) concerning the International System of Units, has provoked discussions on certain of its aspects,

declares,

1. the base units, the supplementary units, and the derived units, of the International System of Units, which form a coherent set, are denoted by the name 'SI units';

2. the prefixes adopted by the CGPM for the formation of decimal multiples and sub-multiples of SI units are called 'SI prefixes';

and *recommends*

3. the use of SI units, and of their decimal multiples and sub-multiples whose names are formed by means of SI prefixes.

Note The name 'supplementary units', appearing in Resolution 12 of the 11th CGPM (and in the present Recommendation) is given to SI units for which the General Conference declines to state whether they are base units or derived units.

14th CGPM, 1971

pascal; siemens ***Pascal; siemens***

The 14th CGPM adopted the special names 'pascal' (symbol Pa), for the SI unit newton per square metre, and 'siemens' (symbol S), for the SI unit of electric conductance (reciprocal ohm).

TAI ***International Atomic Time; function of CIPM*** (CR, 77)

Resolution 1
The 14th CGPM

considering

that the second, unit of time of the International System of Units, has since 1967 been defined in terms of a natural atomic frequency, and no longer in terms of the time scales provided by astronomical motions,

that the need for an International Atomic Time (TAI) scale is a consequence of the atomic definition of the second,

that several international organizations have ensured and are still successfully ensuring the establishment of time scales based on astronomical motions, particularly thanks to the permanent services of the Bureau International de l'Heure (BIH),

that BIH has started to establish an atomic time scale of recognized quality and proven usefulness,

that the atomic frequency standards for realizing the second have been considered and must continue to be considered by CIPM helped by a Consultative Committee, and that the unit interval of the International Atomic Time scale must be the second realized according to its atomic definition,

that all the competent international scientific organizations and the national laboratories active in this field have expressed the wish that CIPM and CGPM should give a definition of International Atomic Time, and should contribute to the establishment of the International Atomic Time scale,

that the usefulness of International Atomic Time entails close coordination with the time scales based on astronomical motions,

requests CIPM

1. to give a definition of International Atomic Time; [8]

2. to take the necessary steps, in agreement with the international organizations concerned, to ensure that available scientific competence and existing facilities are used in the best possible way to realize the International Atomic Time scale and to satisfy the requirements of users of International Atomic Time.

[8] In anticipation of this request, CIPM has asked the Consultative Committee for the Definition of the Second (CCDS), to prepare a definition of International Atomic Time. This definition, approved by CIPM at its 59th session (October 1970), is as follows:

'International Atomic Time [TAI] is the time reference coordinate established by the Bureau International de l'Heure on the basis of the readings of atomic clocks operating in various establishments in accordance with the definition of the second, the time unit of the International System of Units.'

mole ***SI unit of amount of substance (mole)*** (CR, 78)

Resolution 3
The 14th CGPM

considering the advice of the International Union of Pure and Applied Physics, of the International Union of Pure and Applied Chemistry, and of the International Organization for Standardization, concerning the need to define a unit of amount of substance,

decides

1. The mole is the amount of substance of a system which contains as many elementary entities as there are atoms in 0.012 kilogram of carbon 12; its symbol is 'mol'.

2. When the mole is used, the elementary entities must be specified and may be atoms, molecules, ions, electrons, other particles, or specified groups of such particles.

3. The mole is a base unit of the International System of Units.

15th CGPM, 1975

UTC ***Coordinated Universal Time*** (CR, 104)

Resolution 5
The 15th CGPM

considering that the system called 'Coordinated Universal Time' (UTC) is widely used, that it is broadcast in most radio transmissions of time signals, that this wide diffusion makes available to the users not only frequency standards but also International Atomic Time and an approximation to Universal Time (or, if one prefers, mean solar time),

notes that this Coordinated Universal Time provides the basis of civil time, the use of which is legal in most countries,

judges that this usage can be strongly endorsed.

becquerel; gray ***SI units for ionizing radiations*** (CR, 105)

Resolutions 8 and 9
The 15th CGPM

by reason of the pressing requirement, expressed by the International Commission on Radiation Units and Measurements (ICRU), to extend the use of the International System of Units to radiological research and applications,

by reason of the need to make as easy as possible the use of the units for non-specialists,

taking into consideration also the grave risk of errors in therapeutic work,

adopts the following special name for the SI unit of activity: *becquerel*, symbol Bq, equal to one reciprocal second	Resolution 8
adopts the following special name for an SI unit in the field of ionizing radiation: *gray*, symbol Gy, equal to one joule per kilogram[9]	Resolution 9

peta; exa ***SI prefixes peta and exa*** (CR, 106)

Resolution 10
The 15th CGPM

decides to add to the list of SI prefixes to be used for multiples, which was adopted by the 11th CGPM, Resolution 12, paragraph 3, the two following prefixes:

Multiplying factor	*Prefix*	*Symbol*
10^{15}	peta	P
10^{18}	exa	E

(9) Note. The gray is the SI unit of absorbed dose. In the field of ionizing radiation the gray may also be used with other physical quantities also expressed in joules per kilogram; the CCU is made responsible for studying this matter in collaboration with the competent international organizations.

Appendix II Practical realization of the definitions of some important units

1. Length The following recommendation was adopted by the CIPM in 1960 to specify the characteristics of the discharge lamp radiating the standard line of krypton 86:

In accordance with paragraph 1 of Resolution 7 adopted by the 11th CGPM (October 1960) the CIPM recommends that the line of krypton 86 adopted as primary standard of length be realized by means of a hot cathode discharge lamp containing krypton 86 of purity not less than 99 % in sufficient quantity to ensure the presence of solid krypton at a temperature of 64 °K. The lamp shall have a capillary of internal diameter 2 to 4 millimetres, and wall thickness approximately 1 millimetre.

It is considered that, provided the conditions listed below are satisfied, the wavelength of the radiation emitted by the positive column is equal to the wavelength corresponding to the transition between the unperturbed levels to within 1 in 10^8:

1. the capillary is observed end-on in a direction such that the light rays used travel from the cathode end to the anode end;
2. the lower part of the lamp, including the capillary, is immersed in a bath maintained to within 1 degree of the temperature of the triple point of nitrogen;
3. the current density in the capillary is 0.3 ± 0.1 ampere per square centimetre.

(*Procès-Verbaux CIPM*, 1960, **28**, 71; *Comptes rendus 11th CGPM* 1960, 85)

The ancillary apparatus comprises the stabilized current supply for the lamp, a vacuum-tight cryostat, a thermometer for use in the region of 63 K, a vacuum pump, and either a monochromator, to isolate the line, or special interference filters.

Other lines of krypton 86 and several lines of mercury 198 and of cadmium 114 are recommended as secondary standards (*Procès-Verbaux CIPM*, 1963, **31**, 26, Recommendation 1, and *Comptes rendus 12th CGPM*, 1964, 18).

Two monochromatic radiations, one in the visible, the other in the infrared spectral region, produced by helium-neon lasers stabilized on a saturated absorption line of iodine or of methane, are recommended as wavelength standards with the following values (*Procès-Verbaux CIPM*, 1973, **41**, 112)

Line	*Wavelength in vacuum*
Methane, P(7), band ν_3	$3\,392\,231.40 \times 10^{-12}$ m
Iodine 127, R(127), band 11–5 component i	$632\,991.399 \times 10^{-12}$ m

These lines are reproducible with an uncertainty of the order of 1 in 10^{10}; the value of their wavelength in metres is subject to the uncertainty of the standard (the wavelength of the ^{86}Kr line) estimated to be 4 in 10^9. By measuring the beat frequencies of neighbouring lines (for example various components of the hyperfine multiplet of iodine), very exact values of the wavelength differences are obtained.

The wavelength of all these lines varies with pressure, temperature, and composition of the air in which the light travels; the refractive index of the air must therefore in general be measured *in situ*.

To measure end or line standards these radiations are used in an interference comparator, a complicated instrument with mechanical, optical interference, and thermometric components.

The wavelength of the methane line mentioned above multiplied by its frequency (measured by comparison with the ^{133}Cs transition of the definition of the second) yields the speed of propagation of electromagnetic waves in vacuum $c = 299\,792\,458$ m/s, recommended by the 15th CGPM (Resolution 2). This value of c will probably be kept unaltered in the future.

2. Mass The primary standard of the unit of mass is the international prototype of the kilogram kept at the BIPM. The mass of 1 kg secondary standards of platinum-iridium or of stainless steel is compared with the mass of the prototype by means of balances whose precision can reach 1 in 10^8 or better.

By an easy operation a series of masses can be standardized to obtain multiples and sub-multiples of the kilogram.

3. Time Some laboratories are able to make the equipment required to produce electric oscillations at the frequency of vibration of the atom of caesium 133 which defines the second. This equipment includes a quartz oscillator, frequency multipliers and synthesizers, a klystron, phase-sensitive detectors, an apparatus for producing an atomic beam of caesium in vacuum, cavity resonators, uniform and non-uniform magnetic fields, and an ion detector.

Complete assemblies to produce this frequency are also commercially available.

By division it is possible to obtain pulses at the desired frequencies, for instance 1 Hz, 1 kHz, etc.

In the best equipments, the stability and accuracy correspond to an uncertainty of 1 in 10^{12} or even 1 in 10^{13}.

Radio stations broadcast waves whose frequencies are known to about the same accuracy.

There are other standards besides the caesium beam, among them the hydrogen maser, rubidium clocks, quartz frequency standards and clocks, etc. Their frequency is controlled by comparison with a caesium standard, either directly or by means of radio transmissions.

Most time signals broadcast by radio waves are given in a time scale called Coordinated Universal Time (UTC) as recommended by the 15th CGPM (Resolution 5) in 1975. UTC is defined in such a manner that it differs from International Atomic Time (TAI) (1) by an exact whole number of seconds. The difference UTC-TAI was set equal to –10 s starting the first of January 1972, the date of application of the reformulation of UTC which previously involved a frequency offset; this difference can be modified by 1 second by the use of a positive or negative leap second at the end of a month of UTC, preferably in the first instance at the end of December and of June, and in the second instance at the end of March and of September, in order to keep UTC in agreement with the time defined by the rotation of the Earth with an approximation better than 0.9 s (2). Furthermore, the legal times of most countries are offset by a whole number of hours (time zones and 'summer' time).

4. Electrical quantities

So-called 'absolute' electrical measurements, i.e. those that realize the unit according to its definition, can be undertaken only by laboratories enjoying exceptional facilities.

Electric current is obtained in amperes by measuring the force between two coils, of measurable shape and size, that carry the current.

The ohm, the farad and the henry are accurately linked by impedance measurements at a known frequency, and may be determined in absolute value by calculation
(1) of the self inductance of a coil, or the mutual inductance of two coils, in terms of their linear dimensions, or
(2) of the change in electric capacitance of a capacitor in terms of the change in length of its electrodes (method of Thompson-Lampard).
The volt is deduced from the ampere and the ohm.

The uncertainty of these absolute measurements is a few in 10^8 (farad), and a few in 10^6 (ampere).

The results of absolute measurements are obtained by means of secondary standards which are, for instance:

1. coils of manganin wire for resistance standards;

(1) See in Appendix 1, page 39, the definition of TAI given by CIPM at the request of the 14th CGPM (1971, Resolution 1).
(2) The difference UTC–TAI became –16 s on the 1st January 1977.

2. galvanic cells with cadmium sulphate electrolyte for standards of electromotive force;

3. capacitors for standards of electric capacitance (of 10 pF for example).

Application of recent techniques also provides means of checking the stability of the secondary standards which maintain the electric units: measurement of the gyromagnetic ratio of the proton γ_p' for the ampere, measurement of the ratio h/e by the Josephson effect for the volt.

5. Temperature

Absolute measurements of temperature in accordance with the definition of the unit of thermodynamic temperature, the kelvin, are related to thermodynamics, for example by the gas thermometer.

At 273.16 K accuracy is of the order of 1 in 10^6, but it is not as good at higher and at lower temperatures.

The International Practical Temperature Scale of 1968, amended edition of 1975, adopted by the 15th CGPM agrees with the best thermodynamic results to date. The text on this scale is published in *Comité Consultatif de Thermométrie*, 10th session 1974, Annexe T31, and *Comptes rendus 15th CGPM*, 1975, Annexe 2; the English translation is published in *Metrologia*, **12**, 7, 1976.

The instruments employed to measure temperatures in the International Scale are the platinum resistance thermometer, the platinum-10% rhodium/platinum thermocouple and the monochromatic optical pyrometer. These instruments are calibrated at a number of reproducible temperatures, called 'defining fixed points', the values of which are assigned by agreement.

6. Amount of substance

All quantitative results of chemical analysis or of dosages can be expressed in moles, in other words in units of amount of substance of the constituent particles. The principle of physical measurements based on the definition of this unit is explained below.

The simplest case is that of a sample of a pure substance that is considered to be formed of atoms; call X the chemical symbol of these atoms. A mole of atoms X contains by definition as many atoms as there are ^{12}C atoms in 0.012 kilogram of carbon 12. As neither the mass $m(^{12}C)$ of an atom of carbon 12 nor the mass $m(X)$ of an atom X can be measured accurately, we use the ratio of these masses, $m(X)/m(^{12}C)$, which can be accurately determined[(3)]. The mass corresponding to 1 mole of X is then $[m(X)/m(^{12}C)] \times 0.012$ kg, which is expressed by saying that the molar mass $M(X)$ of X (quotient of mass by amount of substance) is

$$M(X) = [m(X)/m(^{12}C)] \times 0.012 \text{ kg/mol}.$$

(3) There are many methods of measuring this ratio, the most direct one being by the mass spectrograph.

For example, the atom of fluorine ^{19}F and the atom of carbon ^{12}C have masses which are in the ratio 18.9984/12. The molar mass of the molecular gas F_2 is:

$$M(F_2) = \frac{2 \times 18.9984}{12} \times 0.012 \text{ kg/mol} = 0.037\,996\,8 \text{ kg/mol}$$

The amount of substance corresponding to a given mass of gas F_2, 0.05 kg for example, is :

$$\frac{0.05 \text{ kg}}{0.037\,996\,8 \text{ kg·mol}^{-1}} = 1.315\,90 \text{ mol}$$

In the case of a pure substance that is supposed made up of molecules B, which are combinations of atoms, X, Y, . . . according to the chemical formula $B = X_\alpha Y_\beta \ldots$, the mass of one molecule is $m(B) = \alpha m(X) + \beta m(Y) + \ldots$.

This mass is not known with accuracy but the ratio $m(B)/m(^{12}C)$ can be determined accurately. The molar mass of a molecular substance B is then

$$M(B) = \frac{m(B)}{m(^{12}C)} \times 0.012 \text{ kg/mol}$$

$$= \left(\alpha \frac{m(X)}{m(^{12}C)} + \beta \frac{m(Y)}{m(^{12}C)} + \ldots \right) \times 0.012 \text{ kg/mol}$$

The same procedure is used in the more general case when the composition of the substance B is specified as $X_\alpha Y_\beta \ldots$ even if $\alpha, \beta, \ldots$ are not integers. If we denote the mass ratios $m(X)/m(^{12}C)$, $m(Y)/m(^{12}C)$, . . . by $r(X)$, $r(Y)$, . . . , the molar mass of the substance B is given by the formula:

$$M(B) = [\alpha r(X) + \beta r(Y) + \ldots] \times 0.012 \text{ kg/mol.}$$

There are other methods based on the laws of physics and physical chemistry for measuring amounts of substance; three examples are given below.

With perfect gases, 1 mole of particles of any gas occupies the same volume at a temperature T and a pressure p (approximately 0.022 4 m^3 at $T = 273.16$ K and $p = 101\,325$ Pa); hence a method of measuring the ratio of amounts of substance for any two gases (the corrections to apply if the gases are not perfect are well known).

For quantitative electrolytic reactions the ratio of amounts of substance can be obtained by measuring quantities of electricity. For example, 1 mole of Ag and 1 mole of (1/2) Cu are deposited on a cathode by the same quantity of electricity (approximately 96 487 C).

Application of the laws of Raoult is yet another method of determining ratios of amounts of substance in extremely dilute solutions.

7. Photometric quantities

Absolute photometric measurements by comparison with the luminance of a black body at the temperature of freezing platinum can only be undertaken by a few well-equipped laboratories. The accuracy of these measurements is somewhat better than 1 %.

The results of these measurements are maintained by means of incandescent lamps fed with d.c. in a specified manner. These lamps constitute standards of luminous intensity and of luminous flux.

The method approved by CIPM in 1937 (*Procès-Verbaux CIPM*, **18**, 237) for determining the value of photometric quantities for luminous sources having a colour other than that of the primary standard, utilizes a procedure taking account of the 'spectral luminous efficiencies' $V(\lambda)$. By its Recommendation CI 1 (1972), CIPM recommends the use of the $V(\lambda)$ values adopted by the International Commission on Illumination (CIE) in 1971.[4] The weighting function $V(\lambda)$ was obtained for photopic vision, i.e. for retinas adapted to light. For retinas adapted to darkness, another function $V'(\lambda)$ gives the spectral luminous efficiency for scotopic vision (CIE 1951); this function $V'(\lambda)$ was ratified by the CIPM in September 1976.

Photometric quantities are thereby defined in purely physical terms as quantities proportional to the sum or integral of a spectral power distribution, weighted according to a specified function of wavelength.

(4) CIE Publications No. 18 (1970), page 43, and No. 15 (1971), page 93; *Procès-Verbaux CIPM*, **40**, 1972, Annexe 1. The $V(\lambda)[= \bar{y}(\lambda)]$ values are given for wavelengths in 1 nm steps from 360 to 830 nm; they are an improvement on the values in 10 nm steps adopted by CIPM in 1933, and previously by CIE in 1924.

Appendix III Organs of the Metre Convention BIPM, CIPM, CGPM

The *International Bureau of Weights and Measures* (BIPM) was set up by the *Metre Convention* signed in Paris on 20 May 1875 by seventeen States during the final session of the Diplomatic Conference of the Metre. This Convention was amended in 1921.

BIPM has its headquarters near Paris, in the grounds (43.520 m^2) of the Pavillon de Breteuil (Parc de Saint-Cloud), placed at its disposal by the French Government; its upkeep is financed jointly by the Member States of the Metre Convention[(1)].

The task of BIPM is to ensure worldwide unification of physical measurements; it is responsible for:

establishing the fundamental standards and scales for measurement of the principal physical quantities and maintaining the international prototypes;

carrying out comparisons of national and international standards;

ensuring the co-ordination of corresponding measuring techniques;

carrying out and co-ordinating the determinations relating to the fundamental physical constants.

BIPM operates under the exclusive supervision of the *International Committee of Weights and Measures* (CIPM), which itself comes under the authority of the *General Conference of Weights and Measures* (CGPM).

The General Conference consists of delegates from all the Member States of the Metre Convention and meets at least once every six years. At each meeting it receives the Report of the International Committee on the work accomplished, and it is responsible for:

discussing and instigating the arrangements required to ensure the propagation and improvement of the International System of Units (SI), which is the modern form of the metric system;

(1) As at 31 December 1976, forty-four States were members of this Convention: Argentina, Australia, Austria, Belgium, Brazil, Bulgaria, Cameroon, Canada, Chile, Czechoslovakia, Denmark, Dominican Republic, Egypt, Finland, France, German Democratic Rep., Germany (Federal Rep. of), Hungary, India, Indonesia, Iran, Ireland, Italy, Japan, Korea, Mexico, the Netherlands, Norway, Pakistan, Poland, Portugal, Rumania, Spain, South Africa, Sweden, Switzerland, Thailand, Turkey, USSR, United Kingdom, USA, Uruguay, Venezuela, Yugoslavia.

confirming the results of new fundamental metrological determinations and the various scientific resolutions of international scope;

adopting the important decisions concerning the organization and development of BIPM.

The International Committee consists of eighteen members each belonging to a different State; it meets at least once every two years. The officers of this Committee issue an *Annual Report* on the administrative and financial position of BIPM to the Governments of the Member States of the Metre Convention.

The activities of BIPM, which in the beginning were limited to the measurements of length and mass and to metrological studies in relation to these quantities, have been extended to standards of measurement for electricity (1927), photometry (1937) and ionizing radiations (1960). To this end the original laboratories, built in 1876–1878, were enlarged in 1929 and two new buildings were constructed in 1963–1964 for the ionizing radiation laboratories. Some thirty physicists and technicians work in the laboratories of BIPM. They do metrological research, and also undertake measurement and certification of material standards of the above-mentioned quantities. BIPM's annual budget is of the order of 5 000 000 gold francs, approximately 2 000 000 US dollars (in 1977).

In view of the extension of the work entrusted to BIPM, CIPM has set up since 1927, under the name of *Consultative Committees*, bodies designed to provide it with information on matters which it refers to them for study and advice. These Consultative Committees, which may form temporary or permanent 'Working Groups' to study special subjects, are responsible for co-ordinating the international work carried out in their respective fields and proposing recommendations concerning the amendments to be made to the definitions and values of units. In order to ensure worldwide uniformity in units of measurement the International Committee accordingly acts directly or submits proposals for sanction by the General Conference.

The Consultative Committees have common regulations (*Procès-Verbaux CIPM*, 1963, **31**, 97). Each consultative Committee, the chairman of which is normally a member of CIPM, is composed of a delegate from each of the major Metrology Laboratories and specialized Institutes, a list of which is drawn up by CIPM, of individual members also appointed by CIPM, and of a representative of BIPM. These Committees hold their meetings at irregular intervals; at present there are seven of them in existence:

1. The *Consultative Committee for Electricity* (CCE), set up in 1927.
2. The *Consultative Committee for Photometry and Radiometry* (CCPR), new name given in 1971 to the *Consultative Committee for Photometry* (CCP), set up in 1933 (between 1930 and 1933 the preceding Committee (CCE) dealt with matters concerning Photometry).

3. The *Consultative Committee for Thermometry* (CCT), set up in 1937.
4. The *Consultative Committee for the Definition of the Metre* (CCDM), set up in 1952.
5. The *Consultative Committee for the Definition of the Second* (CCDS), set up in 1956.
6. The *Consultative Committee for the Standards of Measurement of Ionizing Radiations* (CCEMRI), set up in 1958. In 1969 this Consultative Committee split into four sections: Section I (X and γ rays, electrons); Section II (Measurement of radionuclides); Section III (Neutron measurements); Section IV (α-energy standards). In 1975 this last section was dissolved and Section II made responsible for its field of activity.
7. The *Consultative Committee for Units* (CCU), set up in 1964.

The proceedings of the General Conference, the International Committee, the Consultative Committees and the International Bureau are published under the auspices of the latter in the following series:

Comptes rendus des séances de la Conférence Générale des Poids et Mesures;

Procès-Verbaux des séances du Comité International des Poids et Mesures;

Sessions des Comités Consultatifs;

Recueil de Travaux du Bureau International des Poids et Mesures (this compilation brings together articles published in scientific and technical journals and books, as well as certain work published in the form of duplicated reports).

From time to time BIPM publishes a report on the development of the Metric System (SI) throughout the world, entitled *Les récents progrès du Système Métrique.*

The collection of the *Travaux et Mémoires du Bureau International des Poids et Mesures* (22 volumes published between 1881 and 1966) ceased in 1966 by a decision of CIPM.

Since 1965 the international journal *Metrologia*, edited under the auspices of CIPM, has published articles on the more important work on scientific metrology carried out throughout the world, on the improvement in measuring methods and standards, on units, etc, as well as reports concerning the activities, decisions and recommendations of the organs of the Metre Convention.

Index

(Numbers in italics indicate the pages where the definitions of the units are to be found)